Alexa Echo Instructions

1000+ Alexa Tips and Tricks
How to use your Amazon Alexa Devices

Paul O. Garten

Alexa Echo Instructions

1000+ Alexa Tips and Tricks
How to use your Amazon Alexa Devices

Paul O. Garten

Contents

Introduction

The Amazon virtual assistant can process thousands of commands but how many of these commands do you know? How do you get to build commands? Of course, the Amazon virtual assistant is not a human and can only understands what has been programmed for her to understand. As a result, it behooves you to learn what she can understand and not talk to her like she's a human.

Inside this book, you will learn the patterns of forming a command and many examples of commands that you can use covering a range of categories—media controls, reading books, communication, funs, games, photos, music, video, news and information, sport, productivity and more.

Important notes about using this Book

- This book assumes that you've done all set up necessary for your Alexa-enabled smart devices to respond to voice commands. In other words, it doesn't cover how to set up your smart devices.

- While Alexa may be the household name for the Amazon virtual assistant, you can also call her Amazon, Computer or Echo. You may choose to go with Alexa. We'll be using "Echo" throughout this book.

- Words in **...Italics...** can be changed, but such change must still fall in line with the request. For example, a command such as "Echo, show ...*front door*..." where "front door" is the name of your smart security camera can still be changed to "Echo, show

...*kid's room*..." where "kid's room" is another camera monitoring the kids.

- "Show" or "Watch" implies that you are sending the command to Amazon Echo device with a screen.

- Some commands may not work for you depending on your device and country.

- Forward slash (/) is use in this book for specification and not literarily as seen. Example, "Echo, switch [on / off] kitchen light]. This means you should only say "Echo, switch on the kitchen light" or Echo, switch off the kitchen light" at a time and not the two at once as seen here.

- Sometimes, try a similar word. Example, "Echo, set this photo as <u>wallpaper</u>" instead of "Echo, set this photo as <u>background</u>."

- Some commands may be based on a service subscription and if you don't subscribe to such service, they may not work for you.

1 Getting Started

Volume controls

"Echo, increase the volume."

"Echo, louder."

"Echo, volume down."

"Echo, lower the volume."

"Echo, volume ...*4*..."

"Echo, turn [on / off] whisper mode." Whisper mode helps you to talk to Echo without disturbing others.

Wake Word, Profiles & User Accounts

"Echo, change your name?"

Note: You may not be able to use your devices until the new wake word propagates.

"Echo, switch accounts."

"Echo, which profile is this?"

"Echo, who am I?"

"Echo, what is FreeTime?"

"Echo, speak French."

Bluetooth and WiFi

"Echo, pair."

"Echo, discover."

"Echo, Bluetooth."

"Echo, connect to my phone."

"Echo, disconnect my phone."

"Echo, connect to the internet?"

Note: You must set your phone on pairing mode before trying to pair it with Echo.

2 Book Reading

Audiobooks

"Echo, sign up for Audible."

"Echo, read my books on Audible."

"Echo, show my books on Audible."

"Echo, what's free from Audible?"

"Echo, read ...*Educated*... from Audible."

"Echo, read Girl on the Train from Audible."

"Echo, stop reading in ...*15*... minutes."

"Echo, sign-up for Audible."

"Echo, what popular books are on Audible this week?"

"Echo, what popular books are on Audible today?"

"Echo, read slower."

"Echo, read faster."

Kindle books

"Echo, read my books on Kindle."

"Echo, show my Kindle books."

"Echo, read *Educated* from Kindle."

"Echo, read The Martian from Kindle."

"Echo, sign up for Kindle Unlimited."

Playback controls

"Echo, read my book."

"Echo, next chapter."

"Echo, previous chapter."

"Echo, jump ahead."

"Echo, go back 10 seconds."

"Echo, stop reading in 20 minutes."

"Echo, go ahead 15 seconds."

"Echo, restart the book."

"Echo, restart chapter."

"Echo, pause."

"Echo, resume."

"Echo, stop reading."

3 Communication

Basic controls

"Echo, answer."

"Echo, hang up."

"Echo, [turn on / activate] 'Do Not Disturb.'"

"Echo, [turn off / deactivate] 'Do Not Disturb.'"

Calling & messaging

"Echo, show contact."

"Echo, call Mom."

"Echo, call Paul's mobile phone."

"Echo, call Paul's Echo."

"Echo, turn off video."

"Echo, message John Paul."

"Echo, end call.

"Echo, show my messages."

"Echo, play my messages."

Drop In

"Echo, drop in."

"Echo, drop in on ...*kid's room*..."

"Echo, drop in on Echo."

"Echo, stop drop in."

Announcements

"Echo, announce."

"Echo, announce that I am good."

"Echo, announce that dinner's ready."

"Echo, broadcast."

"Echo, broadcast that breakfast is ready."

"Echo, announce that I'm heading home from the office."

4 Donations

"Echo, donate."

"Echo, make a donation."

"Echo, donate to ...*Doctors Without Borders*..."

5 Live Radio

Try: "Echo play [channel name]."

"Echo, play WNYC."

"Echo, play Z100."

"Echo, how do I play radio stations?"

6 Fun with Echo

Songs

"Echo, sing."

"Echo, sing ...*Christmas carol*..."

"Echo, can you ...*rap*...?"

"Echo, rap for Mom."

"Echo, rap for Dad."

"Echo, sing a spooky song."

"Echo, sing a ...*campfire song*..."

Or

...summer song...

...love song...

...birthday song...

...country song...

...the alphabet...

...a lullaby...

...pirate song...

...the National Anthem... of *...US...*

And more

"Echo, sing [song title]."

"Echo, sing 'Paper Airplanes.'"

"Echo, sing the song that goes, 'let me be the one to love you more'."

Favorites

"Echo, what's your favorite ...*color*...?"

Or

...*sport...?"*

...*book...?"*

...*rock song...?"*

...*superhero...?"*

...*author...?"*

...*word...?"*

...*comic book...?"*

...*animal...?"*

...*game...?"*

...*hat...?"*

...*sport...?"*

...song...?"

...actor...?"

...skateboarder...?"

...hobby...?"

...flower...?"

And more

"Echo, who's your favorite [movie name] character?"

"Echo, who's your favorite *...Merlin...* character?"

...female authors...?"

...baseball team...?"

...baseball player...?"

...poet...?"

And more

Jokes

"Echo, say a joke."

"Echo, tell me a *...math...* joke."

"Echo, tell me a *...science...* joke."

15

Or

...golf...

...football...

...soccer...

...baseball...

...tennis...

...hockey...

...Star Trek...

...Star Wars...

...animal...

...vegetable...

...winter...

...Pokémon...

...ninja...

...dinosaur...

...superhero...

...video game...

...wizard...

...banana...

...chicken...

...coffee...

...dog...
...horse...

7 Photos

Photo commands for Fire Tablets, Echo Spot, Echo Show, and Fire TV

"Echo, show my photos."

"Echo, show my photo albums."

"Echo, next."

"Echo, previous."

"Echo, zoom."

"Echo, zoom in."

"Echo, zoom out."

"Echo, take a photo."

"Echo, take a 4 shot photo."

"Echo, set this photo as a wallpaper."

"Echo, show photos from my [pet / wedding] album."

"Echo, show my This Day Photos."

"Echo, show photos of Mike."

"Echo, take a Sticker photo."

"Echo, set this photo as a background."

"Echo, show photos from last [month / year / weekend]."

"Echo, play a slideshow."

"Echo, pause slideshow."

"Echo, repeat slideshow."

"Echo, turn [on / off] shuffle."

8 Music Audio

Playback

"Echo, shuffle."

"Echo, stop."

"Echo, pause."

"Echo, play."

"Echo, resume."

"Echo, skip."

"Echo, next."

"Echo, repeat the song."

"Echo, what's playing?"

Volume

"Echo, volume up."

"Echo, volume down."

"Alexa, volume 5."

Equalizer

"Alexa, set the bass to 5."

"Alexa, turn up the treble."

"Alexa, set the treble to 6."

"Alexa, reset the equalizer."

Amazon Music Services

"Echo, play the song of the day."

"Echo, try Amazon Music Unlimited."

"Echo, play the ...*Crossfit workout*... playlist from ...*Amazon Prime Music*..."

"Echo, play ...*Because you loved me*... by ...*Celine Dion*... from ...*Amazon Prime Music*..."

"Echo, play ...*One heart*... album by ...*Celine Dion*... from ...*My Music Library*..."

"Echo, play ...*Celine Dion*... from ...*Amazon Prime Music*..."

"Echo, play ...*Top Pop*... station from ...*Amazon Music*..."

Playing from other Music Providers

These include Tidal, iHeartRadio, Spotify, Pandora, SiriusXM, TuneIn, Deezer, Apple Music, and more. To set up any of these music services, go to Settings → Music & Books → Music.

"Echo, play [song title / station / artist name / album / playlist / genre] from [music provider]."

"Echo, play ...*Kiss FM*... on ...*iHeartRadio*..."

"Alexa, play / stop music on .... *iHeartRadio*..."

"Echo, play ...*Drunk*... by ...*Ed Sheeran*... on ...*Spotify*... on ...*music everywhere*..." where "music everywhere" is your Echo speaker group for a music-room music setup.

Customized

"Echo, play songs I listened to last ...*Sunday*..."

"Echo, play some music I heard ...*yesterday*..."

"Echo, add this song to my playlist."

"Echo, create a playlist called ...*Hot*..."

"Echo, add this song to my ...*Hot*... playlist."

Genre, location, era, mood, latest song or album

"Echo, play music from the ...*90s*..."

"Echo, play today's top hits in UK."

"Echo, play happy music."

"Echo, play music from ...*1988*..."

"Echo, play the new ...*Chris Brown*... album."

Lyrics

"Echo, play the song that goes ...*love is all that matters*..."

"Echo, block explicit lyrics."

"Echo, show the lyrics of this song."

Music activities

"Echo, play music for ...*cooking*..."

Other activities could be cleaning, bedtime, yoga, meditation, relaxing, dancing, and more.

"Echo, play ...*rock*... music for a party."

"Echo, play chill music."

"Echo, play upbeat music."

Recommendations

"Echo, play music."

"Echo, play new music."

"Echo, play more music like this."

"Echo, play something similar to ...*Celine Dion*..."

"Echo, enable Today in Music."

"Echo, play Weekly One."

9 Smart Home

Note: Make sure that your concerned smart devices are paired with Echo before attempting these requests.

Surveillance

"Echo, show [event / live video feed] from [camera name]."

"Echo, show live video feed from the ...*front door*..."

"Echo, show the ...*front door*... camera."

Light / Plugs

"Echo, switch [on / off] ...*living room*..."

"Echo, turn the security lights [on / off]," where "security lights" is your smart bulbs group name. You can also create a smart group for plugs.

"Echo, turn on / off the ...*TV plug*..."

Thermostat

"Echo, what's the temperature of the ...*Thermostat*...?"

"Echo, set the ...*House*... to ...*70*... degrees."

"Echo, reduce the temperature of the ...*Thermostat*... to ...*60*... degrees."

"Echo, ...*reduce*... the temperature of the ...*Thermostat*... by ...*2 degrees*..."

"Echo, ...*increase*... the temperature of the ...*House*... by ...*4 degrees*..."

"Echo, set the ...*Thermostat*... to [auto / heat / cool]."

"Echo, set the ...*Thermostat*... to ...*Off*..."

Microwave

"Echo, set the Microwave for ...*5*... minutes."

"Echo, set the Microwave for ...*15*... minutes ...*40*... seconds."

"Echo, add ...*2*... minutes to the Microwave."

"Echo, defrost for ...*15*... minutes"

"Echo, microwave for ...*15*... minutes on [high / low / medium] power."

Multi-room music

"Echo, play music everywhere."

"Echo, play my ...*Hot*... playlist ...*downstairs*..."

"Echo, play ...*jazz*... in the ...*kitchen*..."

"Echo, rename [smart device name]."

"Echo, rename the bedroom plug."

"Echo rename the TV plug."

"Echo, add [name of smart device] to [name of smart device group]."

"Echo, add ...balcony light... to security light group."

10 News & Information

Flash Briefing

To get started, go to Menu → Settings → Flash Briefing in your Alexa app. Some popular Flash Briefings skills include; NPR, Bloomberg, CBNC, Wall Street Journal, Fox News, Washington Post, BBC News, CNN, Reuters, MTV UK News, etc.

Others include Ask Wxbrad (for weather information), Digg (for curated news), Curiosity Daily (for science and technology updates), Marketplace (for news in economics), Daily Tech Headlines (for tech news), Fox Sports (for latest sport news), etc.

"Echo, what's my Flash Briefing?"

"Echo, what are my [news / sport / weather, etc.] Flash Briefings?"

"Echo, play my flash briefings from [flash briefing source / skill name]."

"Echo, play my flash briefings from CNBC."

"Echo, play the news."

Controls

"Echo, [next / previous / cancel]."

"Echo, [pause / stop / resume / continue / next / previous]."

Animals and Planets

"Echo, what's the tallest animal?"

Learn about planets and animals. In future, it may extend to other things.

"Echo, teach me about [type of planet / animal]."

"Echo, teach me about Mars."

"Echo, teach me about Lions."

Books

"Echo, who wrote ...*The Power of Now*...?"

"Echo, tell me about ...*Dan Brown*..."

"Echo, what is ...*Catch-22*...?"

"Echo, when was ...*Educated*... published?"

"Echo, how many books has ...*Michelle Obama*... written?"

"Echo, what's the latest book by ...*Oprah Winfrey*...?"

"Echo, tell me about ...*Napoleon Hill*..."

"Echo, tell me a ...*motivational*... quote."

Business

"Echo, when was ...*Amazon*... founded?"

"Echo, who's the CEO of ...*Walmart*... ?"

"Echo, what's the stock price of General Motors?"

Maths

"Echo, what's 20 times 5?"

"Echo, what's 20 plus 5?"

"Echo, what's 20 minus 5?"

"Echo, what's 20 divided by 5?"

You can also enable the 1-2-3 Math skill for your kids to learn Math even faster. It tests one's ability to add, subtract, multiply, divide, compare, etc. It works in three modes: easy, medium & hard. You may need a Calculator to meet up with allotted time. To enable this skill, say,

"Alexa, enable one two three."

To start using it, "Echo, open one two three."

To have Alexa repeat the question again, say, "Alexa, [say again / repeat],"

To get clearer instructions, say, "Alexa, help."

To get your scores, say, "Alexa, score."

To change game level, say, "Alexa, change the level to [easy / medium / hard]."

Spelling

"Echo, spell the word, magnificent."

Conversions

"Echo, how many ...*centimeters*... are in a ...*meter*...?"

"Echo, what's ...*534*... ...*dollars*... in ...*Pounds*...?"

Dates

"Echo, when is ...*Thanksgiving Day*... in the ...*US*..."

"Echo, what's today's date?"

"Echo, what day is December 25?"

"Echo, how many days do we have until ...*Halloween*...?"

Definitions

"Echo, define ...*Atom*..."

"Echo, how do you define ...*an Atom*...?"

Facts

"Echo, give me an animal fact."

"Echo, tell me some facts."

"Echo, tell me some interesting things."

"Echo, open Random Facts skill."

"Echo, tell Random Facts to give me a fact about ...*Disney*..."

Other categories include, money, weather, Dinosaur, number, food, today, world, etc. You can also enable additional extras with a premium subscription.

Sport updates

"Echo, when next is ...*Chelsea*... football club playing?"

"Echo, what's the score in the game between ...*Chelsea and Manchester city football club*...?

"Echo, what does the ...*English*... premier league table look like?"

Local search

Note: Set your location right before using these commands.

"Echo, which [restaurant / business] is close to me."

"Echo, what top-rated [restaurants / businesses] are close to me?"

"Echo, find the address of [restaurants / businesses] close to me."

"Echo, find the phone number of a [restaurant / business] close to me."

"Echo, find the hours of a [restaurant / business] close to me."

"Echo, what ...*Chinese restaurants*... are nearby?"

"Echo, find a coffee shop around me?"

"Echo, where can I buy a ...*bread*...?"

Language translation

You can translate between French, Spanish, German, Japanese, Italian, Chinese, Polish, Hindi, Portuguese, Dutch, Korean, Danish, Norwegian, Russian, Swedish, Turkish, Romanian, Danish, Icelandic, and Welsh using the Amazon virtual assistant. To get started, say,

"Echo, say ...*goodbye*... in ...*French*..."

"Echo, how do you say ...*come here*... in ...*Spanish*...?

Traffic

Set your traffic preference in the Alexa app to get traffic information within your location. Go to Settings → Alexa Preferences → Traffic to get started.

"Echo, how's traffic?"

"Echo, how is the traffic right now?"

"Echo, what is my commute?"

Weather

"Echo, what's the weather?"

"Echo, is it going to rain today?"

"Echo, how is the weather this weekend?"

"Echo, tell me about the weather in London"

"Echo, tell me about tomorrow's weather."

"Echo, what's the temperature?"

"Echo, what's the weather forecast for ...*Tuesday*...?"

"Echo, what's the humidity today?"

"Echo, what's the weather forecast this weekend?"

Wikipedia

"Echo, Wikipedia Wikipedia ...*Albert Einstein*..."

Notifications

"Echo, what did I miss?"

"Echo, play my notifications?"

"Echo, next."

"Echo, previous."

"Echo, delete my notifications."

Trending News and Pop Culture

"Echo, what's trending?"

"Echo, tell me some weird stuff."

"Echo, what three things do I need to know now?"

Holidays

"Echo, when is the next holiday?"

"Echo, tell me a holiday limerick."

"Echo, why do we celebrate [holiday name]?"

"Echo, how old is Santa Claus?"

"Echo, is Santa Claus real?"

"Echo, where does Santa Claus live?"

"Echo, where is Santa?"

"Echo, track Santa."

"Echo, sing a Christmas carol."

"Echo, tell me a snowman joke."

"Echo, what's your favorite holiday movie?"

"Echo, what are the top holiday movies?"

News

Now, you can get top news from providers such as NPR, CNN, Fox News, Bloomberg, ESPN, Newsy and CNBC.

Try: "Echo, play the news from [news provider]."
"Echo, play news from Fox News."
"Echo, play news from Bloomberg."
"Echo, play news."
"Echo, what are my headlines."
"Echo, skip."
"Echo, next."
"Echo, stop."

11 Podcasts

"Echo, play a podcast."

"Echo, play [title of podcast]."

"Echo, play a podcast."

"Echo, play 'The Drop Out Apple podcast."

"Echo, play Root of Evil Apple podcast."

"Echo, play Root of Evil latest Apple podcast."

"Echo, resume playing Root of Evil Apple podcast."

"Echo, fast forward / rewind 30 seconds."

"Echo, next episode."

Recommendations by categories

Economics: Freakonomics.

Educational: Radio Lab, 99% Invisible.

Finance: Planet Money.

History: Stuff You Missed in History, Revisionist History.

Kids: Story Pirates, Wow in the World.

News: The Daily podcast, Wait Wait...Don't Tell Me!

Science: Science VS.

Self-improvement: Tim Ferris.

Technology: This Week in Tech.

More podcast inspirations

You can also try The Drop Out, Conan O'Brien Needs A Friend, Root of Evil, The Ballad of Billy Balls, Blackout, Dolly Parton's America, To Live and Die in LA, 1619, 13 Minutes to the Moon, The Daily, Overheard at National Geographic, and more.

12 Productivity

Join a conference call

Putting in place or setting up a conferencing solution such as Amazon Chime, Skype, Zoom, or Cisco WebEx set up, and also linking your calendar such as Google G-Suite or Gmail, Apple iCloud, Outlook, or Microsoft Office 365 to the Alexa application, you can join a conference call with others. Set your calendar in the Alexa app via, **Menu → Settings → Calendar & Email.** To get started with a conferencing, say,

"Echo, join my meeting."

"Echo, join my ...*Board meeting...*"

While the conference call is still going on, you can say,

"Echo, hang up" — end the call.

"Echo, press 5" — enter touch tones.

Remember, you must have added the meeting to your calendar before this time.

Running late for a meeting? Let the attendees know about it. Say,

"Echo, I am running ...*10 minutes*... late."

"Running late for ...Board meeting...?" Echo asks.

You reply: "Yes."

Echo then send an email to the attendees with the message, "I am running ...*10 minutes*... late.

You want to go late to the next meeting? Say,

"Echo, I will be ...*10 minutes*... late for my next meeting."

To schedule a 1:1 meeting with a co-worker, say, "Echo, schedule a meeting with Garten" to get started.

To reschedule a meeting, say, "Echo, move / reschedule my ...*Board meeting*... to ...*6 PM*... ...*tomorrow*... / on ...*Thursday*..."

Alarms
"Echo, set an alarm."

"Echo, set an alarm for ...*5 AM*..."

"Echo, set a music alarm."

"Echo, wake me up at ...*2 AM*... to ...*Drunk*... by ...*Ed Sheeran*..."

"Echo, set up an alarm for every ...*Sunday*... at ...*5 AM*..."

"Echo, wake me up to ...*Urban FM*... at ...*4 AM*... on ...*TuneIn*..."

"Echo, set a repeating alarm for weekdays at ...*4 AM*..."

"Echo, what alarms are there?"

"Echo, snooze."

"Echo, cancel my alarm."

"Echo, stop the alarm."

"Echo, change my alarm sound."

"Echo, when is my next alarm?"

"Echo, cancel my alarm for ...*4 AM*..."

"Echo, set up an alarm for ...*5 AM*... every morning with my bedside lamp."

Note: This can only work with compatible smart lights.

Calendars

The Amazon virtual assistant supports Apple (Calendar only), Google (Email and Calendar), Microsoft (Email and Calendar) and Microsoft Exchange (Calendar only) through Office 365 and Outlook.

To get started with Calendar, in the Alexa app, go to **Settings**, scroll down and tap on **Email & Calendar**. Select your preferred service and **Connect Your Account** supplying your login credentials (if necessary). Grant Alexa necessary permissions and finally, your account email and calendar are added.

"Echo, show my calendar."

"Echo, what event do I have on my calendar today?"

"Echo, what's on my calendar at ...*2 PM*... ...*today*...?"

"Echo, what appointment do I have on my calendar on ...*Tuesday*...?

"Echo, what appointment do I have on my calendar on ...*January 10*...?

"Echo, when is my next meeting?"

"Echo, add ...*optometrist appointment*... to my calendar."

"Echo, schedule ...*dinner*... with ...*Dad*... ...*tomorrow*... at ...*2 PM*..."

"Echo, move ...*optometrist appointment*... from ...*2 PM*... to ...*4 PM*..."

"Echo, add ...*meeting*... to my calendar by ...*2 AM*... ...*tomorrow*..."

"Echo, add ...*travel*... to my calendar by ...*5 AM*... on ...*Monday*..."

"Echo, move ...*optometrist appointment*... from ...*2 PM*... on ...*Tuesday*... to ...*10 AM*... on ...*Wednesday*..."

Reminders

"Echo, remind me to ...*visit the dentist*... on ...*Saturday*... at ...*2 PM*..."

"Echo, remind me to ...*join my meeting*... at ...*3 PM*..."

"Echo, remind me to ...*take the laundry out*... in ...*1 hour*..."

"Echo, remind me to ...*walk the dog*... on ...*Saturday*... at ...*6 PM*..."

"Echo, cancel ...*take the laundry out*... reminder."

"Alexa, add a reminder about ...*going to morning mass*... at ...*6 AM tomorrow*..."

"Echo, what reminders do I have?

Timers

"Echo, set a timer for ...*10 minutes*..."

"Echo, set a second timer for ...*30 minutes*..."

"Echo, set a ...*microwave timer*... for ...*8 minutes*..."

"Echo, set a ...*sleep timer...* for ...*1 hour...*"

"Echo, what time is left on the ...*microwave timer...*?"

"Echo, what are my timers?"

"Echo, cancel my timer."

"Echo, cancel /stop my ...*microwave timer...*"

"Echo, stop playing in ...*50 minutes...*"

"Echo, cancel the ...*30 minute...* timer."

Kitchen help

"Echo, suggest a recipe for ...*dinner...*"

"Echo, find / search for a recipe for ...*cupcakes...*"

"Echo, find / search for quick ...*lunch...* recipes."

"Echo, order ...*milk...*"

"Echo, add ...*eggs...* to my shopping list."

"Echo, enable ...*Allrecipes*..."

"Echo, ask ...*AllRecipes*... to show me how to make ...*cupcakes*..."

"Alexa, ask ...*MyChef*... what I can make for ...*dinner*..."

"Echo, play recipe video."

Other recipe skills / services include; SideChef, Food Network, Recipedia, Good Housekeeping, MyChef, OurGroceries, Instant Pot, Meal Idea, Food52, and more.

Lists

"Echo, create a list."

"Alexa, create a ...*books to read*... list."

"Echo, add ...*sugar*... to my ...*shopping list*..."

"Echo, put / add ...*call the carpenter*... on my ...*to-do list*..."

"Echo, what's on my ...*books to read*... list?"

"Echo, remove *...call the carpenter...* from my *...to-do...* list."

"Echo, remove item number *...3...* in my *...shopping...* list."

"Echo, show my *...shopping...* list."

"Echo, clear my *...to-do...* list."

Remember this

"Echo, what can you remember?"

"Echo, remember that *...Mom...* favorite *...color...* is *...orange...*"

13 Shop Amazon

"Echo, track my stuff?"

"Echo, what are my notifications?"

"Echo, search for *...children toys...*"

"Echo, is *...Apple AirPods...* on my *...shopping list...?*"

"Echo, how much is *...Dove cream...?*"

"Echo, order *...Dove cream...*"

"Echo, order *...5...* pieces of *...Dove cream...*"

"Echo, reorder *...Dove cream...*"

"Echo, buy move *..Apple AirPods...*"

"Echo, order *...Dove cream ...black...* color."

"Echo, add *...Dove cream...* to my cart."

"Echo, order everything on my cart."

"Echo, what's in my cart?"

"Echo, cancel my order." (Within 30 minutes after placing order).

"Echo, what's the best-selling ...*children toy*...?"

"Echo, find a best-selling ...*children toy*..."

"Echo, what are your deals?"

Shop Fresh, Prime Now & Whole Foods

"Echo, add ...*eggs*... to my Fresh cart."

"Echo, add ...*beer*... to my Prime Now cart."

"Echo, add ...*sugar*... to my Whole Foods cart."

"Echo, shop Whole Foods."

14 Privacy

"Echo, what did you hear?"

"Echo, delete everything I have just said."

"Echo, delete what I said ...*yesterday*..."

15 Videos, Movies & TV Shows

"Echo, turn [off / on] captions."

"Echo, [play / stop / pause / rewind / fast forward]."

"Echo, [go forward / go backward] by ...*60*... [seconds / minutes]."

"Echo, [restart / play next / next episode / skip back / volume up / volume down]."

"Echo, [switch channel to / change channel to] on CNBC / NBC]."

"Echo, show me [movie title] on [Echo device name]."

"Echo, show me ...*Homecoming*... on ...*Echo Show 8*..."

"Echo, is ...*Homecoming*... any good?"

"Echo, which theater is playing / showing ...*Homecoming*... tonight?"
"Echo, show more."

"Echo, play a movie starring ...*John Boyega*..."

"Echo, show me the closest movie times."

"Echo, who directed the ...*Star Wars*... movie?"

"Alexa, is ...*Star Wars*... any good?"

"Echo, when was ...*The Avengers*... movie released?"

"Echo, what movies are showing tonight?"

"Echo, tell me channels that are showing the Major League Baseball."

"Echo, tell me about soccer broadcast schedule this Saturday."

"Echo, which channel is showing ...*Chelsea vs. Manchester City*... match this Saturday?

"Echo, play / show music videos."

"Echo, show music videos by ...*Ed Sheeran*..."

"Echo, play the ...*No Brainer*... music video."

"Echo, play ...*hip-hop*... music videos."

"Echo, show my video library"

"Echo, show my Watch List."

"Echo, show me [movie / content title]"

"Echo, find [TV series name]."

"Echo, show [genre] movies."

"Echo, show me movies directed by ...*Paul Shaffer*..."

"Echo, enable Stream Player."

"Echo, tell <u>Stream Player</u> to launch ...*Bloomberg*..."

"Echo, tell <u>Stream Player</u> to [play / show / launch] [channel name]."

Tip: Follow link to see Stream Player channels.

"Echo, enable Vevo."

"Echo, play some music videos on Vevo."

"Echo, ask Vevo to play ...*Britney Spears*... music videos."

"Echo, ask Vevo to play ...*Cheap Thrills*... music videos."

"Echo, ask Vevo to play ...*hip-hop*... music videos."

"Echo, ask Vevo to play music videos"

"Alexa, open YouTube."

To watch videos from third party providers such as Hulu, CNBC, or NBC, to Settings → TV & Video in the Alexa app to get started. Once you've completed for any of these video services. You can begin to make request.

"Echo, ask Hulu to tune to ...*ESPN*..."

"Echo, ask Hulu to play ...*Seinfeld*..."

"Echo, show me channels on ...*Hulu*..."

"Echo, rewind."

"Echo, play next episode."

"Echo, play episodes of ...*Shameless*..."

"Echo, change channel to ...*Family Guy*..."

"Echo, ask ...*CNBC*... for pricing info for [stocks / ETFs / futures / indices]."

"Echo, ask ...*CNBC*... for latest news."

"Echo, ask ...*CNBC*... what's happening with the ...*US*... markets?"

"Echo, [play / stream / watch / find / search / search for / show me / open / turn on / tune to] [Video content name] (e.g., Cartoon Network, Channels, Networks, Sci-Fi shows, Movies, Movies with..., Episodes of..., Sports, etc.) on [Hulu / NBC]."

"Echo, search for ...*comedies*... on ...*NBC*..."

"Echo, show ...*Bring the funny*... on ...*NBC*..."

"Echo, show the trailer of the movie ...*Star Wars*..."

FireTV special

"Echo, watch [show / movie] title on FireTV"

"Echo, play [show / movie] title on [app name] on FireTV."

"Echo, play [movie genre] on FireTV."

"Echo, play [movie genre] on [app name] on FireTV."

"Echo, find TV shows and movies on FireTV."

"Alexa, search for [title / genre] on FireTV."

"Echo, show movie titles with [actor's name] on FireTV."

"Echo, find [TV show / movie] on [app name] on FireTV."

"Echo, search for [genre / title] on [app name] on FireTV."

"Echo, show me [TV shows / movies] on [app name] on FireTV."

"Echo, go to [network / channel] on [app name]."

"Echo, show [channel / network] on FireTV."

For Prime members using Echo devices with a screen.

"Echo, stream football."

"Echo, watch Thursday night football."

"Echo, play the Seahawks game."

Play Amazon Original movie

"Echo, play [movie name]."
"Echo, play Photograph."
"Echo, play Peterloo."
"Echo, play Late Night."

Play Amazon Original series

"Echo, play Comicstaan S2."

Others include, Kung Fu Panda: The Paws of Destiny 1B, Dino Dana S3, The Boys S1, Niko and the Sword of Light, Goliath, and more.

The show and tell feature for the vision impaired.

If you are having difficulty identifying an item, you can hold the item closer to the Echo Show camera and then ask Echo to identify it.

Try:
"Echo, can you identify what I am holding?"

16 Skills

"Echo, enable / disable [skill name]."

"Echo, enable ...*Harry Potter*..."

"Echo, disable ...*Food Network Kitchen*..."

"Echo, open MLB."

Link your MLB account to get updates about your favorite games and also watch highlights if you are using an Echo device with a screen.

Get started

"Echo, how do I enable / disable a skill?"

"Echo, what top skills do you have?"

"Echo, what popular skills do you have?"

"Echo, show me skills."

"Echo, recommend a skill for me?"

"Echo, recommend a ...*productivity*... skill for me?"

"Echo, what ...*finance*... skill should I enable."

"What ...*fitness*... skill should I enable?"

Useful skills to enable

Note: Read more about a skill before enabling it for your device.

Finance

Capital One, Opening Bell, TD Ameritrade, Cryptocurrency, etc.

Productivity

Quick Events, Giant Spoon.

Smart home & car

Yonomi, Automatic, Harmony, Anova Culinary, Joule, etc.

Food and Drink

MySomm, What beer, The Bartender, Meat Thermometer, Meal Idea, Pizza Hut, etc.

Fitness

7-Minute Workout, 5-Minute Plank Workout, Fitbit, Track by Nutritionix, Guided Meditation, etc.

Weather

Big Sky, Feels Like, Fast Weather, etc.

Travel

Kayak, Uber, Lyft, etc.

Entertainment

Valossa Movie Finder, This Day in History, Radio Mystery Theater, Short Bedtime Story, Ambient Noise, Reddit TIL, etc.

Podcast and radio

AnyPod, Stitcher, Learn Something Radio, TED Talks. Games: Trainer Tips, The Wayne Investigation, The Magic Door, Earplay, Potterhead Quiz, etc.

Educational

This Day in History, NASA Mars, Geo Quiz, Spelling Game, Ultimate History Quiz, etc.

Family & Kids

Amazon Storytime, Kids Court, Yes Sire, Freeze Dancers, SpongeBob, Animal Game, etc.

Fun & Games

My Pet Rock, Song Quiz, Jeopardy, Sports Jeopardy, Teen Jeopardy, Twenty Questions, Heads Up, Categories Game, Pet Rock, The Music Quiz, etc.

News & entertainment

NPR, CNBC, The Tonight Show with Jimmy Fallon, Daily Show, AnyPod, TuneIn Live, etc.

Meditation & Relaxation

Headspace, Rain Sounds, Beach Sounds, Sleep Sounds, Beach Video, Fireplace Video, etc. "Echo, help me relax."

17 Skill Blueprints

"Echo, how do I create a skill?"

"Echo, enable / open [your skill name]."

18 Routine

"Echo, when I say 'It's evening," turn on the security light."

You are now setting up a routine using your voice.

Printed in Great Britain
by Amazon